BASIC GUIDE BOOK TO MIG VS TIG WELDING

Everything you need to know about mig and tig from scratch

Troy Kelsey

Table of Contents

CHAPTER ONE

INTRODUCTION

It's always amazing to see a perfect TIG weld on a product. A stack of dimes looks like a perfect TIG weld. An indicator of the level of skill of the welder is the accuracy and evenness of the weld beads. However, an almost perfect MIG weld on a product like a truck frame is a sign of the strength and efficiency that can be achieved by MIG welding. Two of the most popular types of welding in many industries are MIG and TIG welding. MIG and TIG welding (or Metal Inert Gas [MIG] and Tungsten Inert Gas [TIG]) are used in many industries, including automotive, nuclear, marine, aerospace, oil, and nuclear. Understanding the basics of MIG and TIG welding is important. You can select the right technology for your application if you have a better grasp of the basics.

Understanding the differences and similarities

between MIG and TIG welding will help you to understand them.

MIG Welding, also known as Gas Metal Arc Welding or (GMAW),

This is a type of welding measurement in which the welding bend occurs between the consumable wires and the workpiece. The consumable wire is melted by the welding arc. The consumable wire is fed through the handle, which the user controls using a trigger. The consumable wire is fed to the weld pool. It can be built to the required detail depending on the machine setting. Carbon dioxide is the most popular shielding gas used for MIG welding. It is more affordable than helium or argon, and this is mainly because of its low cost.

TIG Welding, also known as Gas Tungsten Arc Welding or GTAW (Tungsten Inert Gas),

This type of welding involves the formation of a welding

arc between a tungsten electrode and the workpiece. The welding torch contains a collet that holds the tungsten electrode. The user should hold the tungsten electrode approximately one-eighth of an inch from the workpiece. This is necessary to ensure that the arc does not form a weld puddle. You can weld with or sans a consumable filler rod, which is manually dipped in the weld pool.

Because the TIG welding process is manual controlled,

skill and technique are crucial to creating a visually appealing bead. Fusion TIG welding doesn't require filler. It is a simple technique that involves waving the torch across the workpieces to be joined. To produce a weld using filler rod, you need to pulse the foot pedal. The filler rod is dipped into the weld pool with one hand while the torch is held in the other. This requires coordination between the operator's feet, both hands and eyes.

Similarities mig and tig

There are many similarities between MIG and TIG welding processes. There are several common elements between the MIG and TIG welding processes, including welding current, deposition of welding material, heat-affected zones, and applications. Both MIG and TIG can use DC current or electrode negative electrical currents. TIG can also use AC current to create an aluminum-optimal weld puddle. MIG welding uses current that flows through the welding wire, creating arcs between it and the workpiece. TIG welding also uses an electrical current that passes through the tungsten electrode. The arcs are from the electrode to your workpiece.

The MIG and TIG welding processes are similar to conventional welding. This is the creation of huge heat by using an electric arc. This is how you can create a controlled puddle from liquid metal to join two metals together. To develop the weld bead, both MIG and TIG welding can use filler materials. Filler material improves joint strength. MIG welding uses filler material to increase strength by automatically feeding the consumable wire in the weld pool. The filler rod is used to build up the weld in TIG welding. The filler rod is inserted into the weld pool.

The Heat Affected Area

This is the area of the workpiece that is most affected by high-temperature welding. It can be identical between TIG and MIG processes. . Micro welding and advanced TIG techniques are exceptions. The MIG handle can be pulsed to control the heat affected area. This is a way that design engineers might request. This effectively stops and starts the weld beam. This pulsing strategy can be used by automotive repair workers to weld metal sheets onto vehicles.

TIG welding is much more manageable and can be used to control the heat affected area. The welder's foot pedal controls the intensity of the arcs. MIG and TIG welding are used in many common areas, including automotive sheet metal repair, automobile racing components, and pipe fitting.

CHAPTER TWO

MIG AND TIG DIFFERENCES

MIG and TIG welding share many similarities. However, there are important differences. These include the use of shielding gases, speed, precision, and automation. Both MIG and TIG welding require shielding gasses, although the type of gas used varies. The shielding gas keeps the weld puddle safe from the elements. Because of the oxygen, hydrogen and nitrogen in the atmosphere, the weld can be contaminated by the air around us. Shielding gases can be so crucial for weld purity, that we recommend that welds are performed in an enclosed atmosphere of inert gas. This is highly recommended when welding titanium. Standard MIG welding applications require CO2 gas. The gas is abundant and therefore it is the most common and economical choice for MIG welding.

MIG welding uses helium, oxygen, and argon as shielding gases less often. These gasses are reserved for aluminum and specialty metals. To shield the weld puddle, TIG welding almost always uses 100 per cent argon gas. A mixture of helium and argon is sometimes used, for example, when welding high-nickel content metals. This is what gave rise to the old term heli-arc welding. Mixtures of argon, hydrogen or nitrogen are less common. These mixtures are used to TIG weld certain stainless steels. The speed at which the weld is applied is another difference between MIG welding and TIG welding. MIG welding is a manual process that requires constant application of the welding rod. A MIG welder can apply the weld bead quickly.

High Quality or High Speed

TIG welding is a skilled trade that can produce a weld beam at a rapid pace. A TIG welder can rarely beat MIG welding's speed. This is due to the manual nature and limitations of TIG welding. The MIG process can produce a weld quickly and in a consistent way, making it a common tool in automated welding processes. MIG welding is automated, which contrasts with TIG welding, which requires high precision and manual work. In highly precise applications, where precision is needed, TIG welding may be an advantage. For example, in a helicopter engine or space shuttle exhaust.

History of the Technologies

The Post-War era saw a lot of demand for goods, which led to the development of MIG welding technology. Factories were able produce large quantities of automobiles and ships. MIG welding technology was a major contributor to this development. MIG welding became the most popular welding method after the use of CO2 as a shielding agent in the 1950s. As an alternative to MIG welding, flux core welding was introduced in the 1950's. Flux core welding is a tubular welding process that uses flux material within the wire. This is useful when shielding gas may not be readily available. It is less efficient than MIG because flux core welding doesn't produce as good a weld.

In the 1940's, TIG welding was officially made a welding process. The shielding gas that was used at the time to protect the welding process, helium, gave it its original name "Heliarc". The original process was DC Electrode Positive. The current was then reversed by technological advances, making DC Electrode Negative (DCEN), possible. DCEN was able to weld ferrous materials better, but non-ferrous materials proved more difficult. This was until AC current TIG welding. The high-frequency arc starter was the final major advancement. To start the arc, the operator doesn't need to strike the tungsten electrode on the workpiece. This could often contaminate or damage the electrode.

Let's now look at future applications and equipment after you have been familiarized with both MIG- and TIG welding.

Applications to mig and tig welting from household - industrial

There are many MIG and TIG welding machines that can be used in both residential and industrial settings. There are many differences in the welding applications for a household and an industrial job site. A few examples of a household application would include fixing things like a backyard fence or a lawn care tool. Or perhaps a truck bumper mounts.

As you can see, household repairs need a lighter duty welding machine. Industrial applications are different from household applications. They can be found in shipyards, aircraft or automobile factories, repair shops for aviation or automobiles, and mine sites, nuclear facilities as well as building structures or oil fields. These industrial applications require a much higher level of welder skill than any household application. While a household application requires economy and ease of operation, industrial applications require durability, long duty cycles and an inspection-able final weld bead. It is easy to see why, as human lives can depend on weld quality.

TIG and MIG Welding Equipment, from Residential to Industrial

For the most part, the household user and hobbyist welder require a different welding feature to what an industrial worker would need. A household user will often prefer an intuitive user interface. It is also beneficial to be able to use household electricity. A 220v MIG welding machine is an ideal choice for hobbyists and household users. It comes with a bottle CO_2 shielding gas. The welder plugs into the 220v outlet found in most households. This outlet is often used for dryers or washing machines. This welder is capable of producing clean, good-looking welds for various light duty applications. They are also affordable. This category is very popular with Lincoln Electric, Miller and ESAB.

Industrial Applications

MIG and TIG welding are required for industrial applications. MIG welders for heavy duty use have many features, such as large MIG filler wires, extended duty cycles, heavy-duty components and even a generator to provide power. Welding equipment for industrial use must withstand repeated exposure to harsh environments.

MIG welders are used in pipe fitting, mining, and steel erecting. Welders are often used in areas where conditions aren't ideal. A welding company might be needed to repair an earthmover on an earth moving job site. In this situation, time is of the essence and the fastest way to repair the damage is to use a truck-mounted welder equipped with an integrated generator. TIG welders are also used in aerospace, marine, automotive, and other areas where machines are constantly being used.

TIG welders must be able to withstand constant industrial use. Often, machines are water-cooled. The water cooling is necessary to ensure that they don't overheat or become damaged. For marine applications, aluminum is the preferred material for weight and corrosion. TIG welding or MIG welding is used with the appropriate aluminum spool gun to attach railings, deck plate, or hull reinforcements. A water-cooled TIG torch with water cooling is essential for TIG welding aluminum with AC current. This is because aluminum melts faster.

The Future of TIG Welding

According to an article by aws.org, "Welding Forges into the Future", the industry will continue to grow because welding is the best method to join materials. Technological improvements will continue to make a steady, slow growth. This is especially true in the aluminum welding category, and possibly into the composites category. Emerging markets that are experiencing industrial revolutions will likely drive further growth. There is also the domestic market for welding automation. Further speculations are made in this article, concluding that friction stir welding (or laser beam welding) will continue to be more common in the future.

While friction stir welding (a form of advanced metal joining that NASA has developed) and laser beam welding (which is commonly used to join sheet metals), are both undergoing continuous advancements, the most popular technologies will continue to be MIG or TIG welding. Automation has made MIG welding more efficient in recent years. Automobile factories are at the forefront in automation of assembly line welding. The goal of welding automation is to improve profit margins. However, automation can also improve welding quality because welding robots are not fatigued like human operators. Arc data monitoring could be the future of welding automation. This is the place where a human welder can see problems with the weld pool and make minor adjustments to the weld parameters. These minor adjustments could be made by robots in the future, theoretically.

Do Something

You now have a better understanding of the MIG/TIG welding processes. Now, it's time to get out there and weld! If you have read this far and are interested in learning how to weld with the MIG, TIG, or other processes, it is time to get started! You can also learn more about welding by visiting your local welding shop and speaking to a friendly salesperson. You could also sign up for a welding class at your community college.

There are many ways to learn more about welding and to try it. Next, you should decide which avenue to try and then start your journey.

www.ingramcontent.com/pod-product-compliance
Ingram Content Group UK Ltd.
Pitfield, Milton Keynes, MK11 3LW, UK
UKHW021643190726
13853UKWH00001B/32

9 798758 478790